The Baobab and the Elephant

by Curtis Kotlarczyk

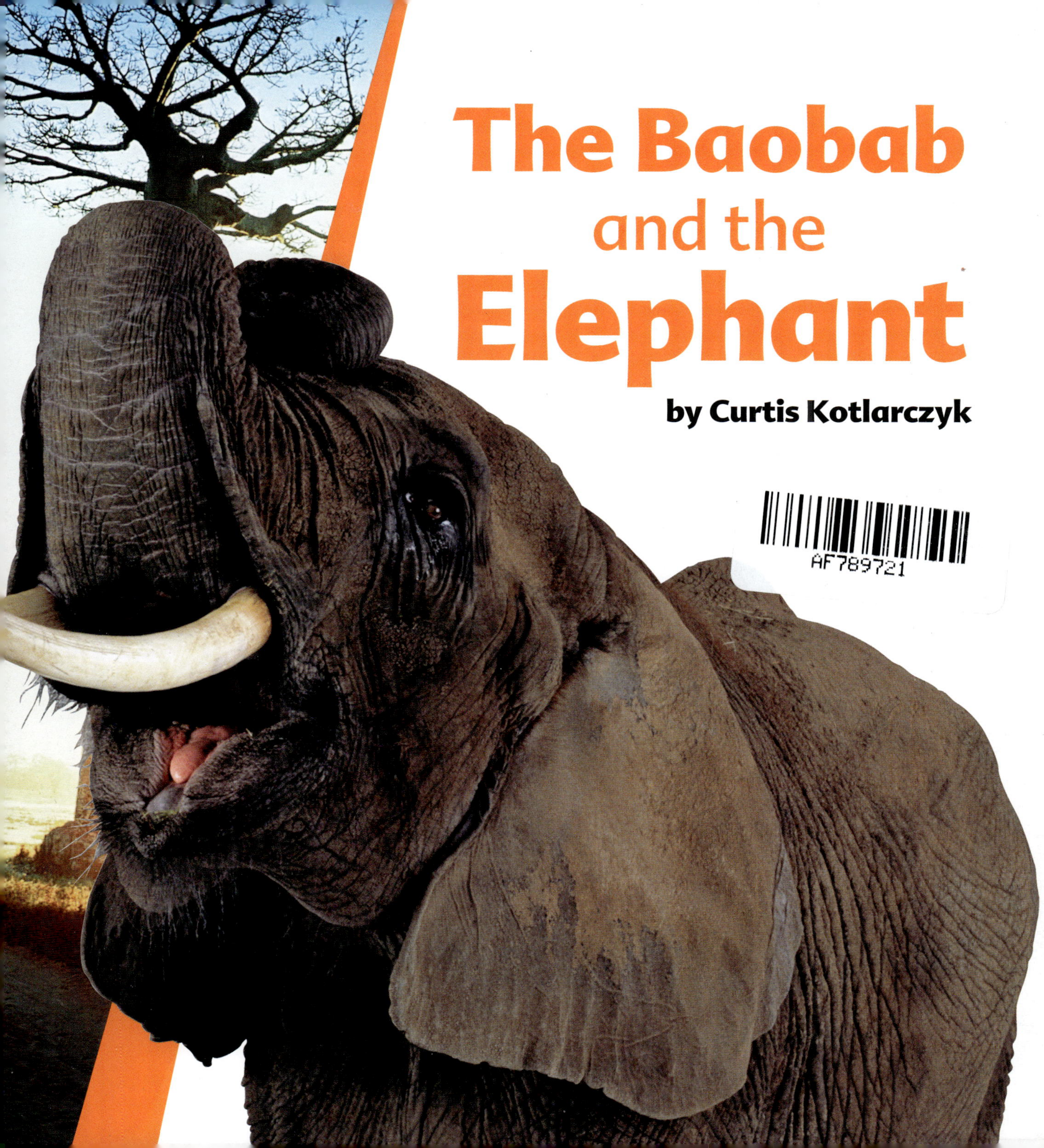

Every living thing has basic needs.

Even two very different living things!

The baobab tree needs
a place to live and grow.
So does the elephant!

The baobab needs air.
So does the elephant!

The baobab needs water.
So does the elephant!

The baobab's roots spread out underground to take in water.

The elephant uses its trunk to lift water to its mouth.

The baobab needs food. So does the elephant!

The baobab makes its own food. It uses light, air, and water to make food in its leaves.

What is the elephant's favorite food?

The baobab!